Understanding the Elements of the Periodic Table™

THE HALOGEN ELEMENTS

Fluorine, Chlorine, Bromine, Iodine, Astatine

Greg Roza

rosen publishing's
rosen
central®

New York

Published in 2010 by The Rosen Publishing Group, Inc.
29 East 21st Street, New York, NY 10010

Library of Congress Cataloging-in-Publication Data

Roza, Greg.
The halogen elements: fluorine, chlorine, bromine, iodine, astatine / Greg Roza.—1st ed.
   p. cm.—(Understanding the elements of the periodic table)
Includes bibliographical references and index.
ISBN 978-1-4358-3556-6 (library binding)
1. Halogens—Popular works. 2. Halogen compounds—Popular works. 3. Periodic law—Popular works. I. Title.
QD165.R69 2010
546'.73—dc22

                                                            2009012539

Manufactured in Malaysia

CPSIA Compliance Information: Batch #TW10YA: For Further Information contact Rosen Publishing, New York,
New York at 1-800-237-9932

On the cover: Illustrations of the halogen atoms, paired with each element's square from the periodic table. Clockwise from top left to right are chlorine, bromine, astatine, and iodine, with fluorine in the center.

Contents

Introduction

Everything in our world is made up of elements—substances such as iron (Fe), gold (Au), oxygen (O), and tin (Sn). When combined, elements form more complex substances called compounds. All of the elements are listed on the periodic table, which allows students and scientists to easily observe how they are similar and how they are different.

The periodic table is made up of rows and columns of elements. The rows of elements are called periods, and the columns are called groups. The elements in a single period or group usually have common characteristics. At the same time, they can be very different.

Group 17 of the periodic table includes (from top to bottom) fluorine (F), chlorine (Cl), bromine (Br), iodine (I), and astatine (At). These elements are called the halogens. They are the most reactive elements on the periodic table, meaning they quickly form compounds with other elements and are rarely, if ever, found isolated in nature. The pure forms of the halogens are all highly poisonous, but their compounds are often very useful. The halogens are also different in several ways. For example, the halogen group is the only one that contains all three states of matter (solids, liquids, and gases) at room temperature.

People have known of and used minerals that contain the halogens for thousands of years. Minerals containing fluorine have long been used to

Developed around 1846 by pioneer photographer John Plumbe, this is one of the first photographs of the U.S. Capitol in Washington, D.C. This type of photograph (called a daguerreotype) required the use of chemicals called silver halides, which contains one or more halogens.

purify metals. Chlorine is a main component of salt, which was first used thousands of years ago to preserve and season foods. Ancient people in the Middle East used a special bromine compound to make an expensive and beautiful purple dye. Some halogens have even been used as currency! Read on to learn more about the amazing halogens.

Chapter One
Meet the Halogens

This chapter will introduce you to each of the halogens, including their basic characteristics, how they were first used, and who discovered them and how. The halogens can be very dangerous because of their high reactivity. However, they are also used to make many useful substances.

Fluorine

Fluorine is a pale yellow gas that is highly corrosive and poisonous in its pure form. It is the most reactive element known to man, and it easily forms compounds with most elements. Fluorine reacts violently when combined with some elements, particularly hydrogen (H). At one time, liquid fluorine and liquid hydrogen were used as rocket fuel. Other fluorine compounds are used to make industrial acids, refrigeration chemicals, and insecticides. Fluorine is used in the enrichment of uranium (U), a highly radioactive element. Enriched uranium is used as fuel in nuclear reactors. A form of fluorine is also added to toothpaste and water supplies to help prevent cavities!

The mineral fluorspar—calcium fluoride (CaF_2)—has been used for thousands of years to help cleanse metals of impurities during smelting. Eventually, scientists began to suspect that fluorspar contained an

A dentist fits a mouthpiece filled with a fluoride treatment into a patient's mouth. Much like the fluoride added to drinking water and toothpaste, this treatment is designed to strengthen the teeth and reduce the risk of cavities.

unidentified element, but it was difficult to isolate. Today, we know that's because this element, fluorine, resists being separated from other elements.

In 1886, French chemist Henri Moissan successfully isolated the element using a process called electrolysis on a solution of potassium hydrogen fluoride (KHF_2) in liquid hydrogen fluoride (HF). Moissan earned the 1906 Nobel Prize in Chemistry for his discovery.

Chlorine

Chlorine is a pale green, strong-smelling gas. The element chlorine is very poisonous, but some of its compounds are very useful. Chlorine is used to

This woman is testing the chlorine levels in a public swimming pool. Chlorine is used to kill bacteria in pool water, and it is generally very safe. However, too much chlorine can irritate the skin and eyes.

make countless industrial and consumer products, including plastics, solvents, cleansers, dyes, and insecticides. An effective water purifier, it is commonly used to kill bacteria in drinking water and swimming pools. It is also used in many cleansers and disinfectants. Bleach is one of the most common products containing chlorine.

The earliest known chlorine compound is sodium chloride (NaCl), also called halite or table salt. German-Swedish chemist Carl Wilhelm Scheele was the first person to isolate pure chlorine in 1774, although he wasn't aware of it. Scheele combined the mineral pyrolusite, or manganese dioxide (MnO_2), with hydrochloric acid (HCl), creating a pale green gas. He thought the gas might contain oxygen. In 1810, English chemist Sir

What Is Electrolysis?

Electrolysis is the use of electricity to change one substance into another. This process can separate some elements from their compounds. Two electrodes are placed in a liquid that contains compounds of several elements, some of which are in the form of ions. Ions are electrically charged atoms or groups of atoms. The positive electrode (anode) attracts negatively charged ions in the liquid. The anode pulls the electrical charge off of the ion, which causes the ion to form a new substance. This new substance can be an element, such as a halogen. The negative electrode (cathode) attracts the positively charged ions. The cathode pushes charge into the ion, which also causes the ion to change into a new substance. These new substances build up on or around the electrodes until enough can be collected. This process is used to purify many elements, including fluorine and chlorine.

Humphry Davy established that the gas was a chemical element and named it chlorine.

Bromine

At room temperature, bromine is a brownish-red, bad-smelling liquid. It is a poisonous substance that can cause irritation when touched, and it can be fatal when inhaled. In compound form, bromine is mainly found in seawater and brine wells. It is used in making flame-retardant materials, agricultural pesticides, water purification chemicals, photographic chemicals, dyes, and medicines.

About three thousand years ago, the ancient Phoenicians used a bromine compound to make a very special purple dye. Tyrian purple

Shown here is the shell of a spiny murex sea snail, the creature from which Tyrian purple dye was once harvested.

(also known as imperial purple) was made from a mucus harvested from a special sea snail. It was very difficult to make and very expensive. Elemental bromine was officially discovered in 1826. That year, Carl Lowig, a German chemistry student at the University of Heidelberg, showed his professor a red, smelly liquid he had isolated from seawater. The professor encouraged Lowig to continue studying the liquid, but Lowig became busy with other projects. The delay gave French chemist Antoine-Jérôme Balard enough time to publish a paper describing the new element, which he named bromine, in 1826. Balard prepared bromine by distilling it from a mixture of seaweed ash and chlorine.

Iodine

Iodine is a black, metallic-looking solid. When gently heated (or exposed to air at room temperature) iodine sublimates, or turns directly from a solid to a gas. Iodine vapor is a stunning violet color. It is also a poison that irritates the eyes, nose, and lungs.

Although iodine is rare in Earth's crust, it is commonly found in seawater. You might be most familiar with the iodine used to disinfect

Iodine is sometimes used as a contrast agent to make certain organs appear brighter in X-rays. This X-ray shows the chest cavity of a person who has had a metal stent inserted in his aorta.

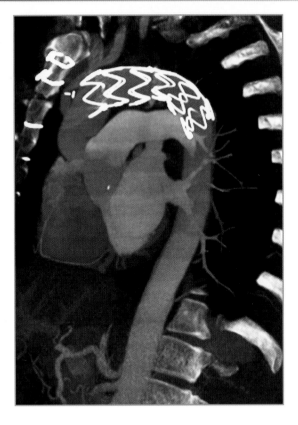

wounds; it was once included in many first-aid kits. It is also used to sterilize water. As an industrial chemical, iodine is used to make dyes, photography chemicals, lightbulbs, and medicines. The human body uses iodine compounds to make certain hormones.

French chemist Bernard Courtois accidentally discovered iodine in 1811. During the Napoleonic Wars, the French army used wood ashes to make gunpowder. When they ran low on wood, they began to burn seaweed instead. Seaweed contains high levels of iodine. Courtois used sulfuric acid to clean the vats needed to make gunpowder. One day he used too much sulfuric acid, which produced a violet gas. The gas condensed on the sides of the vat, forming purple crystals. After running simple tests on the substance, Courtois gave samples of it to other scientists who continued to study it. In 1813, Sir Humphry Davy named the element iodine.

Astatine

Astatine is a highly radioactive element. Its atoms give off a great amount of energy in the form of very tiny particles. The most stable form of astatine has a half-life of only 8.1 hours. Some forms last less than a second!

What's in a Name?

Element	Word Origin	Meaning	Reason for Name
Fluorine	*fluere* (Latin/ French)	flow or flux	Fluorspar was used as a "flux," or a substance used to remove impurities from metals.
Chlorine	*chloros* (Greek)	pale green	Pure chlorine gas is pale green.
Bromine	*bromos* (Greek)	stench	Pure bromine gas smells very bad.
Iodine	*iodes* (Greek)	violet	Iodine vapor is violet.
Astatine	*astatose* (Greek)	restless, unstable	Astatine is a highly unstable element that quickly decays to form elements lower on the periodic table.

When radioactive elements such as uranium and thorium (Th) decay, they form astatine, which in turn quickly decays to form more stable elements. Astatine's instability makes it the rarest naturally occurring element. At any given time, there is only about 1 ounce (28.3 grams) of the element in Earth's crust.

Astatine is also a synthetic element. That means it can be manufactured in a laboratory. Astatine was first synthesized at the University of California, Berkeley, in 1940 by scientists Dale R. Corson, Kenneth R. MacKenzie, and Emilio Segrè. This was accomplished by bombarding an atom of bismuth (Bi) with very small high-energy particles. Three years later, evidence of natural astatine was discovered. Less than a gram of astatine has been created, and most is used for study.

Chapter Two
Atoms, Subatomic Particles, and the Halogens

Did you know that all matter in the universe is made up of unbelievably tiny building blocks called atoms? According to British nuclear physicist and professor Jim Al-Khalili, there are more atoms in a single glass of water than there are glasses of water in Earth's oceans. Although atoms are so small that we cannot see them, even with the aid of powerful microscopes, scientists have discovered a great deal about them over the past one hundred years.

Atoms are made up of even smaller pieces called subatomic particles. At the center of every atom is a cluster of subatomic particles called protons and neutrons. These form the atom's nucleus. Protons have a positive electrical charge, and neutrons have no charge. Electrons orbit the nucleus in different layers, or shells. Each shell can hold a specific number of electrons. Each electron has a negative electrical charge that is attracted to the positive charges of the protons in the nucleus.

The number and combination of these particles give elements their individual characteristics. A minute change in an atom's subatomic particles can change the way it acts; it can even change the type of element it is! Understanding more about protons, neutrons, and electrons will help us to better understand the halogens and what makes them act the way they do.

English chemist John Dalton (1766–1844) is the person generally credited for inventing atomic theory. He came up with the theory after observing how elements always combined in certain ratios.

Atomic Number

The known elements are organized on the periodic table in order of the number of protons in the nucleus of a single atom. The first element on the table, hydrogen, has one proton. Its atomic number is 1. The second element, helium (He), has two protons. Its atomic number is 2. Atomic numbers continue to go up by one with each succeeding element on the periodic table. Of the halogens, fluorine has the lowest atomic number—9. Astatine, with eighty-five protons, has the highest atomic number of the halogens.

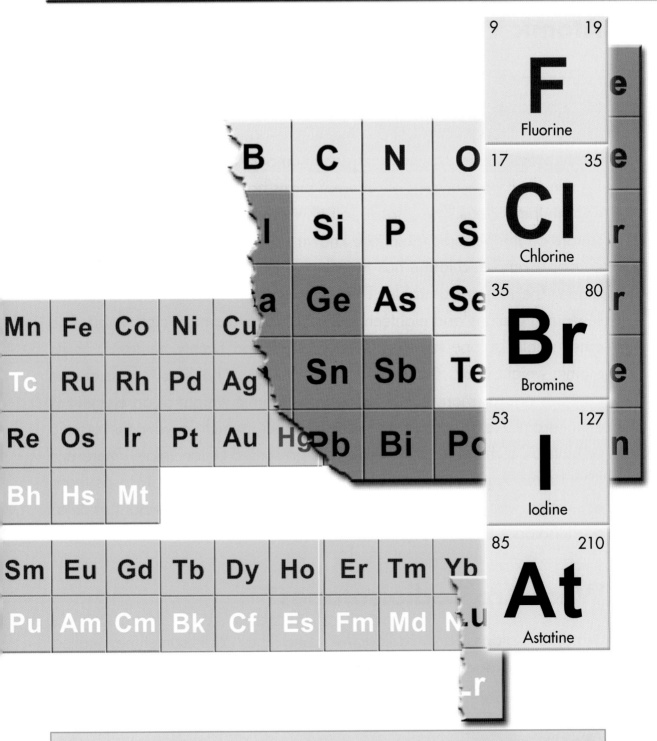

This diagram shows a close-up of the right side of the periodic table, where the halogen group resides. As shown here, each square on the table shows the element's name, symbol, atomic number, and atomic weight.

Atomic Weight and Isotopes

Atomic weight is expressed as atomic mass units (amu). Atomic weight corresponds to the total number of protons and neutrons in the nucleus of an atom. Although atoms of a single element always have the same number of protons, they can have different numbers of neutrons. In fact, some elements can have dozens of different forms based on the number of neutrons in the nucleus. These forms, all of which are represented by a single square on the periodic table, are known as isotopes.

For example, chlorine has twenty-six isotopes, only two of which are stable. The rest are unstable. The most common isotope of chlorine has seventeen protons and eighteen neutrons, so its atomic weight is 35. The other stable isotope has seventeen protons and twenty neutrons, so its atomic weight is 37. Other chlorine isotopes can have as few as eleven neutrons and as many as thirty-four. The atomic weight of an element is an average weight of all its stable isotopes, which is why the atomic weights on some periodic tables are represented by decimals. There are about three times as many chlorine atoms with a weight of 35 as there are with a weight of 37. This is why the average atomic weight of chlorine's isotopes is 35.453 amu.

What Are Radioisotopes?

Unstable isotopes are radioactive, and they are sometimes called radioisotopes. A radioactive element loses energy and mass in the form of subatomic particles. Over time, radioisotopes decay to form a more stable isotope. Some radioisotopes take millions of years to decay; others take just milliseconds. This means that some radioisotopes, and some elements in general, last for less than a second. Radioactive substances can be very dangerous to living things because of the energy they give off, which is called radiation.

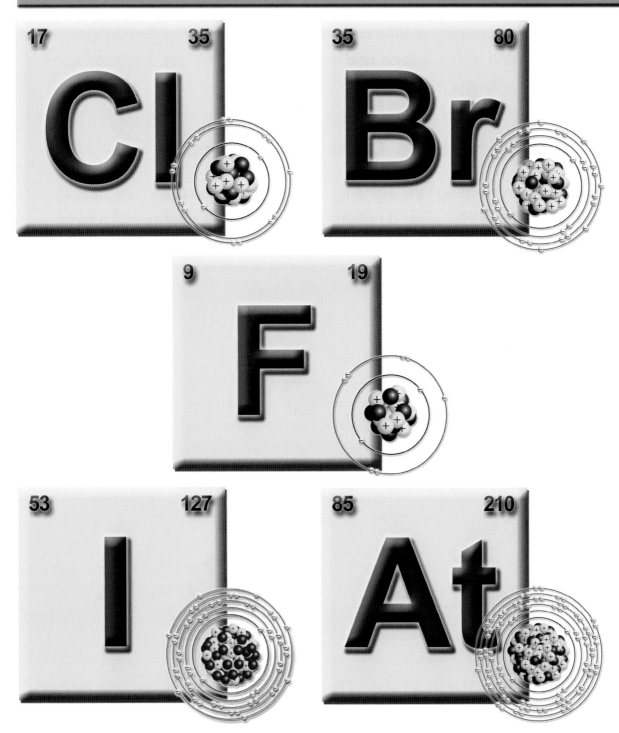

Although no one is completely sure what an atom looks like, scientists use diagrams like the ones shown here to represent them. The red and blue spheres represent neutrons. The yellow spheres are protons. The tiny spheres orbiting the nucleus are electrons.

All isotopes of astatine are radioactive. The rest of the halogens have at least one stable isotope and many unstable isotopes. Most of those isotopes are synthetic and not found in nature.

Electrons and Chemical Bonds

When an atom has the same number of electrons as protons, it is electrically neutral. However, an atom may gain or lose electrons. An atom with

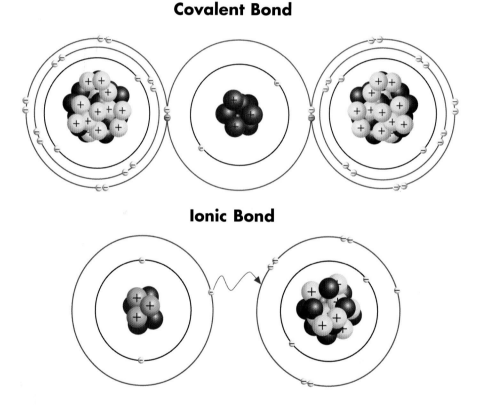

These diagrams illustrate the two most common types of bonds: covalent and ionic. In the top example, the center beryllium atom is sharing the red electrons with the chlorine atoms. In the bottom example, the lithium atom on the left is losing an electron to the stronger fluorine atom on the right, which will cause them to bond together.

a different number of electrons than protons carries an electrical charge. This type of atom is called an ion.

The electrons in the outer shell are called valence electrons. They are capable of forming bonds with other atoms to form molecules. Two of the most common bonds are covalent and ionic bonds. Covalent bonds occur when two atoms share pairs of electrons. Ionic bonds occur when one atom is strong enough to take an electron away from a weaker atom. The resulting ions are attracted to each other and form an ionic bond.

Electronegativity and Diatomic Particles

Electronegativity is the ability of an atom or molecule to attract electrons and form covalent bonds with them. The greater an atom's electronegativity, the more reactive it is. This means it is more likely to form compounds with other atoms and less likely to be found alone in nature.

The halogens are some of the most reactive elements on the periodic table. This is because they all have seven electrons in their outermost electron shells, leaving one open space for another electron. Free halogen atoms constantly try to steal an electron from surrounding atoms to become more stable. This is a process called oxidation. Fluorine is the most electronegative of all the elements. Iodine is the least reactive halogen, although it is more reactive than many other elements.

Pure halogens take the form of a diatomic molecule. Diatomic molecules are particles that contain two atoms of the same or different elements connected with a covalent bond. The symbols for diatomic halogens are: F_2, Cl_2, B_2, I_2, and At_2. Diatomic halogen molecules are far more common in laboratory settings and industrial processes than in nature.

Halogen atoms commonly form ionic bonds with metals to form salts. Take, for example, common table salt, which is made up of a sodium (Na)

The Halogens at a Glance

	Fluorine	Chlorine	Bromine	Iodine	Astatine
Chemical Symbol	F	Cl	Br	I	At
Atomic Number	9	17	35	53	85
Atomic Weight	19	35	80	127	210
Protons	9	17	35	53	85
Neutrons	10	18	45	74	125
Electrons	9	17	35	53	85
Melting Point	−363.32°F (−219.62°C)	−150.7°F (−101.5°C)	19.04°F (−7.2°C)	236.3°F (113.5°C)	575.6°F (302°C)
Boiling Point	−306.62°F (−188.12°C)	−29.27°F (−34.04°C)	137.804°F (58.78°C)	363.2°F (184°C)	638.6°F (337°C)

ion and a chlorine ion. A chlorine atom has room for one electron in its outer shell, and a sodium atom has just one electron in its outer shell. The chlorine atom steals that single electron away from the sodium atom. As a result, the sodium atom becomes a positively charged ion, and the chlorine atom becomes a negatively charged ion. These two atoms are attracted to each other, forming an ionic bond.

Chapter Three
Finding and Refining the Halogens

Industrial countries need raw minerals to fuel their industries. Millions upon millions of dollars are spent and earned in the mining, refining, and sale of industrial chemicals. The halogens—with the exception of astatine—are highly important materials used for countless industrial and commercial applications. In this chapter, you will learn where the halogens are found, how they are gathered, and how they are isolated for shipment and use.

Finding and Purifying Fluorine

Fluorine is the eighteenth most abundant element. It is most commonly found in the glassy mineral fluorspar. Fluorspar is mostly made up of calcium fluoride (CaF_2), but often contains impurities such as the element yttrium (Y). It is found in Earth's crust in many areas of the world, particularly China, Kenya, Mexico, Mongolia, Russia, South America, Switzerland, and the United States. Cryolite (Na_3AlF_6) is another important fluorine mineral that has been found mainly in Greenland.

Today, fluorine is isolated using large-scale electrolysis similar to the process Henri Moissan originally used. This process requires large amounts of electricity. Potassium fluoride and hydrogen fluoride are mixed in a special copper (Cu) container. Graphite electrodes are then

Shown here are blue fluorspar crystals. Fluorspar comes in many different shades, including yellow, purple, and green. The crystals often have more than one color, making them particularly beautiful.

lowered into the mixture. Fluorine gas forms near the anode. At the same time, hydrogen gas forms at the cathode. Great care must be taken during this process because the two gases will explode if mixed. The fluorine gas is compressed until it changes to a liquid. It is then ready to be shipped.

Finding and Isolating Chlorine

Chlorine is the sixteenth most abundant element on Earth. It is found most plentifully in seawater. In fact, after hydrogen and oxygen, it is the most abundant element in seawater, which is about 1.94 percent chlorine.

Chlorine can be collected from salt deposits in and around rivers that have a higher-than-average salt content. White salt flats can be seen in this aerial view of the area where the Ural River meets the Caspian Sea in Kazakhstan.

In the United States, chlorine is often refined from sodium chloride taken from brine wells in Michigan and Arkansas. A brine well is a subterranean pocket of minerals left over after an ancient sea evaporated. The solid minerals are high in salts such as sodium chloride. To collect these minerals, water is pumped deep into Earth's crust. The water mixes with salts to form a mixture called brine. Then the brine is pumped back up to the surface where the salts can be separated and collected.

Chlorine is one of the most important industrial chemicals in the world. It is used for countless products and manufacturing processes. Pure chlorine is produced using electrolysis. There are two main methods of doing this, one of which is the mercury cathode cell process.

During this procedure, brine is placed in a container with graphite anodes. At the bottom of the container is a bed of mercury (Hg) that acts as the cathode. When an electric current passes through the solution, pure chlorine gas collects near the anode, and the sodium mixes in with the mercury. The chlorine gas is collected. The sodium is separated from the mercury and mixed with water to make pure hydrogen gas and sodium hydroxide (NaOH), which is sold to manufacturers.

Extracting and Refining Bromine

Bromine is the sixty-second most abundant element on Earth. The bromine compound used to make Tyrian purple dye was originally harvested from a special type of sea snail found mainly in the Mediterranean Sea. Making even a small amount of dye was a long and difficult process that required thousands of snails. Tyrian purple dye became an important product of trade, along with gold, gems, spices, and textiles. Today, bromine is most often extracted from seawater or brine wells. It is collected the same way chlorine is collected, and often at the same time.

Following the identification of bromine in 1826, bromine production was accomplished by boiling seawater or brine using scrap wood. This process produced solid sodium chloride, which was removed. The resulting liquid, rich in bromine ions, was treated with sulfuric acid (H_2SO_4). The acid extracts bromide from the water by taking an electron away from bromide ions. This produces pure bromine atoms, which combine to form Br_2 molecules.

An American chemical industrialist named Herbert Henry Dow invented the "blowing out" method of refining bromine in 1889. First, he treated brine with chlorine. Chlorine reacts with the bromine ions, removing electrons from them and turning them into bromine atoms and molecules. (The chlorine is changed into chlorine ions.) Burlap bags

The structure in this photo is a mill in Midland, Michigan, where Herbert Henry Dow set up a laboratory to experiment with bromine and other elements. This is where he first produced pure bromine using the "blowing out" method.

soaked in the brine were hung up and air was blown through them, releasing bromine gas. As Dow improved the process, he began using electrolysis to separate the bromine ions instead of using chlorine.

Today, most industrial bromine is produced using the "steaming out" process. First, sulfuric acid is mixed with brine to aid the eventual release of bromine. Next, the solution is poured down the inside of a special tower. Chlorine steam is fed into the bottom of the tower. As the chlorine rises and passes through the brine solution, it converts bromine ions to bromine atoms and releases bromine gas. This is extracted at the top of the tower. The gas is condensed to a liquid, purified, and prepared for shipment.

What Is Fluorescence?

In 1852, mathematician and scientist George Gabriel Stokes discovered that, under the right conditions, some types of fluorspar can be made to glow! Stokes shined invisible ultraviolet light on a sample of fluorspar. The atoms in the fluorspar absorbed the ultraviolet light, which increased their energy. The atoms then released the extra energy to become more stable. Stokes discovered that the energy emitted by the atoms had longer wavelengths than the invisible ultraviolet light they had absorbed. This change in wavelength created visible light, and the fluorspar glowed. Stokes coined the term "fluorescence" (from fluorspar) to describe this phenomenon, which occurs in many other minerals as well.

Finding and Refining Iodine

Iodine is the sixty-fourth most abundant element on Earth. The manner in which iodine is refined depends on where it is collected. The first important commercial sources of iodine were saltpeter beds in Chile. Chilean saltpeter—or sodium nitrate ($NaNO_3$)—contains several iodine compounds.

To separate the iodine, saltpeter is boiled in water and then cooled. This creates saltpeter crystals and a solution rich in iodate ions (IO_3^-). Next, the solution is divided into two parts. Sodium hydrogen sulfate ($NaHSO_4$) is mixed with one part, which changes the iodate ions to iodide ions (I^-). The solutions are combined again. The iodide ions change the iodate ions to iodine.

As with other halogens, iodine is collected from many brine wells around the world. The process used to refine iodine from brine is very

A girl walks in front of a mural in the mining town of Maria Elena in northern Chile. The town is built around a saltpeter mine.

similar to the process used to refine bromine. Seawater could potentially be another source of iodine, but the concentration of iodine in seawater is too low to collect economically.

Some species of seaweed and kelp, however, absorb iodine and store it in their cells. Iodine was once produced by burning seaweed, boiling the ashes in water, and treating the solution with chemicals. Today, seaweed is usually shredded, soaked in water, and filtered to create a liquid high in iodide ions. Similar to the "steaming out" process for bromine, chlorine gas is passed through the solution. The chlorine atoms take electrons from the iodide ions, which form chloride ions and iodine. The solution is boiled until iodine vapors condense at the top of the container.

Chapter Four
Halogen Compounds

A compound is a substance made up of two or more elements bonded together. The resulting substance often displays characteristics different from those of the individual elements.

While halogens by themselves can be very harmful, halogen compounds are often safe and even beneficial for our health. Many pharmaceuticals and anesthetics are made from halogen compounds. A substance to which one or more halogens have been added is referred to as halogenated.

Fluorocarbons and Freons

The most reactive and common of the halogens, fluorine is found in numerous compounds in nature. The only elements it doesn't combine with are the noble gases helium (He) and neon (Ne). Some fluoride compounds are very helpful, while others are deadly.

Fluorocarbons are chains of carbon (C) atoms with fluorine and, sometimes, hydrogen atoms bonded to them. The bond that forms between fluorine and carbon atoms is one of the strongest bonds in all of chemistry. Fluorocarbons form from hydrocarbons, or chains of carbon atoms with hydrogen atoms bonded to them. Both types of molecules are organic compounds, meaning they contain carbon atoms bonded to each

On March 1, 1977, Oregon became the first state to ban the use of aerosol cans that used fluorocarbon propellants. Notices like this one were posted in stores all over the state. Other states soon followed Oregon's lead.

other. A fluorocarbon forms when fluorine replaces some or all of the hydrogen atoms in a hydrocarbon. They are used to make solvents, polymers, anesthetics, and other chemicals.

A type of fluorocarbon that also contains chlorine, called a chlorofluorocarbon (CFC), was once widely used as a coolant in refrigerators and air conditioners and as a propellant in aerosol cans. By the 1970s, scientists discovered that CFCs damage Earth's ozone layer in the upper atmosphere, which protects the planet from the sun's high-energy radiation. As a result of this discovery, CFCs are not as widely used as they once were.

Fire Fighters

Organic compounds that contain some or all of the halogens (except astatine) are used to make halon fire extinguishers. There are many different kinds, but all contain carbon and one or more halogens. Extinguishers that contain bromine are usually the most effective. They are most often used when water-based extinguishers could cause as much damage as a fire, as in a computer facility or museum. Halon-1211 (CF_2ClBr) and halon-1301 (CF_3Br) are the two most commonly used today.

Bromocarbons

Bromine compounds are used to make fire-retardant materials, disinfectants, dyes, and many other applications. At one time, ethylene bromide ($BrCH_2CH_2Br$) was added to leaded gasoline to reduce engine noise.

Bromine easily joins with hydrocarbons to make organic compounds called bromocarbons. Methyl bromide, also called bromomethane (CH_3Br), is one of many effective bromine-based pesticides. However, it is also very dangerous for people, wildlife, and Earth's atmosphere. Rain can wash it into nearby streams and water supplies. Some foods that are not properly cleaned may contain methyl bromide. In addition, scientists have found that methyl bromide damages the ozone layer. Most countries are in the process of phasing out the use of methyl bromide.

Halogenated Plastics

One of the most useful applications for halogens is in the manufacture of plastics. Most halogen-containing plastics are made by combining

This worker is installing PVC pipes in the foundation of a home that is being constructed.

a halogen with a gas called ethylene (C_2H_4), which is a byproduct of the oil industry. Polyethylene is an important polymer made up of a string of ethylene molecules bonded together. Polychloroethylene—also called polyvinyl chloride, PVC, or just vinyl—is an important plastic made from vinyl chloride. Vinyl chloride (C_2H_3Cl) is ethylene with one hydrogen atom replaced by one chlorine atom. PVC is frequently used in the construction industry. It is used to make siding, window frames, plumbing, flooring tiles, and other building materials. It is also used in the production of clothes, upholstery, toys, and swimming pools. PVC is a thermoplastic, which means it can be easily molded when it is hot, although it can be brittle when cool. When mixed with other materials, it becomes highly durable and flexible.

Polytetrafluoroethylene—also called polyvinyl fluoride, or PVF—is a corrosion-resistant halogenated plastic with numerous uses. It is most often used as a protective coating or film. Common uses include vinyl siding for homes and buildings, and a coating for the interior of containers used to transport corrosive materials. You are probably most familiar with the PVF brand name Teflon, a nonstick coating for pans and cooking utensils.

Hydrohalic Acids

Each of the halogens is capable of forming an acid when joined with hydrogen and dissolved in water. These acids are sometimes referred to as hydrohalic acids. You might be familiar with hydrochloric acid, which is hydrogen chloride (HCl) gas dissolved in water. It is a strong and very useful compound with many industrial applications. It is also the liquid in our stomachs that breaks down, or digests, the food we eat.

Hydrofluoric acid—hydrogen fluoride (HF) dissolved in water—is a weak acid used mainly in the production of fluorine compounds. It is also used to etch glass and remove impurities from some metal, such as stainless steel. Hydrobromic acid—hydrogen bromide (HB) dissolved in water—is stronger than hydrochloric acid. It is mainly used in the manufacture of other chemicals. Hydriodic acid—hydrogen iodide (HI) dissolved in water—is even stronger, but it has few applications.

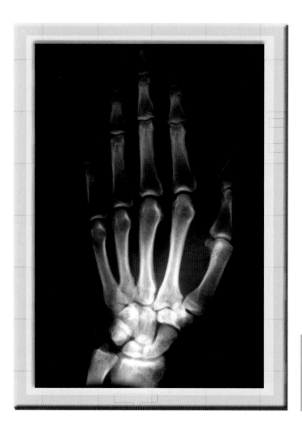

Take a Picture

Silver halides are light-sensitive compounds used to make photographic film. Silver chloride and silver iodide were used as photographic chemicals as early as 1839. Later, silver bromide was

This is an X-ray image of a human hand. Iodide is used in making X-ray film.

found to be more effective. Film is covered with a layer of gelatin that contains tiny silver halide crystals, a mixture that is commonly called an emulsion. When exposed to light, the silver and the halogen separate. The silver solidifies and darkens to create an image.

Silver iodide is still used in the manufacture of film, although digital cameras have reduced its importance in the photography industry. However, it is used for the film used in X-ray imaging. It also has a very different use—to "seed" clouds. Rain falls when water condenses and freezes on dust particles. Silver iodide crystals are similar to ice crystals. When a cloud is "seeded" with silver iodide, water can condense on the crystals, initiating rainfall.

Astatine Compounds

Astatine is chemically similar to iodine and forms salts with metals, but the compounds do not last long. Once again, astatine's instability makes studying its compounds nearly impossible. Scientists believe astatine acts much like other halogens in forming acids. It reacts with hydrogen to form hydrogen astatide (HA+). When dissolved in water, HA+ forms hydroastatic acid.

Chapter Five
Halogens in Our World

In their pure forms, all of the halogens are poisonous and potentially harmful to people. Touching, inhaling, or ingesting them can result in damage to body tissues. Large doses can be fatal. However, the halogens also have benefits to human health.

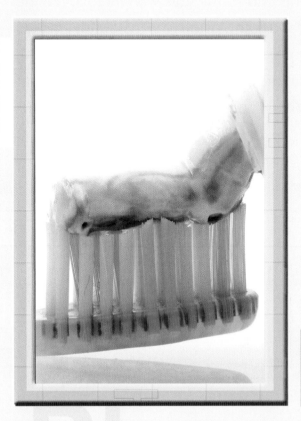

Fluorine's Uses

Centuries before fluorine was identified, the mineral fluorspar was used to purify metals. Today, fluorine compounds are most commonly used in the purification of aluminum. We get a very small amount of fluorine in our daily

Many toothpastes on the market today contain fluoride to help make our teeth stronger.

diets. Fish and shellfish usually have higher concentrations of the element. Traces of fluorine bond with other elements in your bones and teeth. This helps keep them solid and strong. Too much fluorine in your diet, however, can result in tooth decay, osteoporosis, and damage to the kidneys, nerves, and muscles.

Water Fluoridation

Fluoridation is the practice of adding fluoride ions to a water supply for the purpose of improving a population's dental health. After studies conducted in the 1940s, scientists discovered that people who lived in areas where the drinking water was naturally fluoridated had lower cases of tooth decay. Fluoridation of water supplies began in 1945 when the town of Grand Rapids, Michigan, added fluoride to its drinking water. The American Dental Association (ADA) has long been a strong proponent of water fluoridation in the United States. Many dentists recommend that children between the ages of six and sixteen have regular doses of fluoride to strengthen their teeth. Fluoride bonds with the enamel on teeth to form an even stronger material called fluorapatite ($Ca_5(PO_4)_3F$).

Too much fluoride can result in a condition called fluorosis, which is found mainly in children. In mild cases, children develop white patches on their teeth. Worse cases result in brownish and cracking teeth. Regular trips to the dentist can help identify and stop fluorosis.

Chlorine's Uses

Of all the halogens, chlorine is the most widely used. Many commercial and industrial products and processes rely on chlorine, including plastics, textiles, dyes, insecticides, food processing and storage, appliances, building materials, pharmaceuticals, and detergents. Chlorine is effective at killing harmful bacteria, such as *E. coli*. It is used to disinfect drinking water and public pool water. You might be most familiar with chlorine in the form of sodium hypochlorite (NaClO), also called bleach. Thanks to chlorine (and other disinfectants), people today are much healthier than they were years ago.

Chloride and Health

Pure chlorine is a dangerous and potentially deadly element. Chloride ions, on the other hand, are needed to keep our nervous and digestive systems healthy and to aid in the transfer of nutrients across cell walls. Chloride also helps us move our muscles and keep our kidneys healthy. Chloride ions are primarily found in the fluid outside of cells, in the blood, and in stomach fluids. The recommended daily intake of chloride ions is between .026 and .032 ounces (750 and 900 milligrams). We get plenty of chloride from the foods we eat, much of it from table salt.

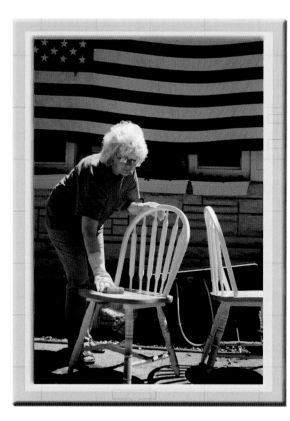

This woman is using a bleach solution to clean chairs. Bleach is an effective disinfectant.

Bromine's Uses

The most common use for bromine is in the manufacture of brominated flame retardants (BFRs), which are chemicals that reduce the chance of an object catching fire. They also slow the rate at which things burn. BFRs are used to make electronics, furniture, rugs, textiles, mattresses, and insulation, to name a few. Also, bromine, chlorine, and fluorine are used to make halon fire extinguishers. They work by removing the source of oxygen from a fire. Bromine is also used to make agricultural chemicals and pesticides, photography chemicals, dyes, and disinfectants. Similar to chlorine, bromine is also used to purify water.

Bromine and Health

Although everyone has a small amount of bromine in their bodies, it is not needed for life. We get bromine from the foods we eat, especially seafood. Too much bromine, however, can cause potentially life-threatening health problems. Scientists have discovered that an abundance of bromine in the body can replace other more important halogens, particularly iodine in the thyroid gland.

Iodine

Tincture of iodine, a solution of iodine in alcohol and water, is an antiseptic that was once commonly included in many first-aid kits. In addition to disinfecting wounds, iodine tablets are used to purify drinking water. One of iodine's most common uses today is in the manufacture of dyes, inks, and stains. It is even used to color bacteria so scientists can tell different kinds apart.

Iodine and Human Health

Of all the halogens, iodine is the most important to human health. It is stored in the thyroid gland and helps regulate metabolic processes. It is the heaviest known element needed by living things to remain alive. Living things use iodine in the form of iodide salt. We get much of the iodine we need from seafood and from vegetables grown in iodine-rich soil.

People who live where the soil may not contain iodine can supplement their diet with iodized salt. A common seasoning in many households and restaurants, iodized salt is table salt with a small amount of potassium iodide, sodium iodide, or iodate. An iodine deficiency can result in thyroid problems, particularly a condition called goiter. Iodine deficiency is particularly dangerous for pregnant women and children.

Possible Uses for Astatine

Because astatine lasts a very short time, very few people have ever had any contact with it. It is also a very dangerous substance. Only trained scientists can synthesize and work with astatine.

Scientists hope that astatine may someday prove helpful in the area of radiation therapy. Due to the fact that astatine does not last long,

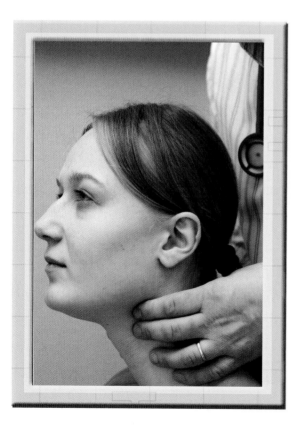

scientists hope to use it as a radioisotope tracer. Like other halogens, astatine concentrates in the thyroid gland. Once there, scientists can use machinery that is sensitive to radioactive particles to make a detailed picture of the gland for examination and diagnosis.

A doctor is checking this woman's thyroid gland, which is located toward the front of the neck below the larynx, or voice box.

Halogen Lights

The halogens are used to make lightbulbs that are more efficient than regular bulbs. In regular bulbs, a tungsten (W) filament is surrounded by nitrogen (N) or argon (Ar) in a sealed glass container. The tungsten gets so hot that it glows. Bulbs create far more heat than they do light. The tungsten gets so hot that it evaporates and collects on the inside of the glass bulb. Eventually, the filament weakens and breaks.

Halogen lights are more efficient than regular bulbs. The space around the tungsten filament contains some halogen gas. As the tungsten evaporates, it joins with the halogen and is deposited back on the filament. To put it simply, the halogen helps recycle the tungsten, allowing the bulb to last longer than regular bulbs. Halogens also operate at a higher temperature and produce whiter light.

Halogens and Our World

From ancient royal dyes to modern medical procedures, the halogens have long been an important part of our world. Our bodies depend on several of the halogens to remain healthy. Scientists continue to experiment with these highly reactive elements. Who knows what future applications the halogens will provide?

The Periodic Table of Elements

Group

IA	IIA	IIIB	IVB	VB	VIB	VIIB	VIIIB	VIIIB
1	2	3	4	5	6	7	8	9

Atomic Number →

9	19
F	
Fluorine	

17	35
Cl	
Chlorine	

35	80
Br	
Bromine	

Name of Element

Period

1

1	1
H	
Hydrogen	

2

3	7
Li	
Lithium	

4	9
Be	
Beryllium	

3

11	23
Na	
Sodium	

12	24
Mg	
Magnesium	

4

19	39		20	40		21	45		22	48		23	51		24	52		25	55		26	56		27	59
K			**Ca**			**Sc**			**Ti**			**V**			**Cr**			**Mn**			**Fe**			**Co**	
Potassium			Calcium			Scandium			Titanium			Vanadium			Chromium			Manganese			Iron			Cobalt	

5

37	85	38	88	39	89	40	91	41	93	42	96	43	98	44	101	45	103
Rb		**Sr**		**Y**		**Zr**		**Nb**		**Mo**		**Tc**		**Ru**		**Rh**	
Rubidium		Strontium		Yttrium		Zirconium		Niobium		Molybdenum		Technetium		Ruthenium		Rhodium	

6

55	133	56	137	57	139	72	178	73	181	74	184	75	186	76	190	77	192
Cs		**Ba**		**La**		**Hf**		**Ta**		**W**		**Re**		**Os**		**Ir**	
Cesium		Barium		Lanthanum		Hafnium		Tantalum		Tungsten		Rhenium		Osmium		Iridium	

7

87	223	88	226	89	227	104	261	105	262	106	266	107	264	108	277	109	268
Fr		**Ra**		**Ac**		**Rf**		**Db**		**Sg**		**Bh**		**Hs**		**Mt**	
Francium		Radium		Actinium		Rutherfordium		Dubnium		Seaborgium		Bohrium		Hassium		Meitnerium	

Lanthanide Series

58	140	59	141	60	144	61	145	62	150	63	152	64	157
Ce		**Pr**		**Nd**		**Pm**		**Sm**		**Eu**		**Gd**	
Cerium		Praseodymium		Neodymium		Promethium		Samarium		Europium		Gadolinium	

Actinide Series

90	232	91	231	92	238	93	237	94	244	95	243	96	247
Th		**Pa**		**U**		**Np**		**Pu**		**Am**		**Cm**	
Thorium		Protactinium		Uranium		Neptunium		Plutonium		Americium		Curium	

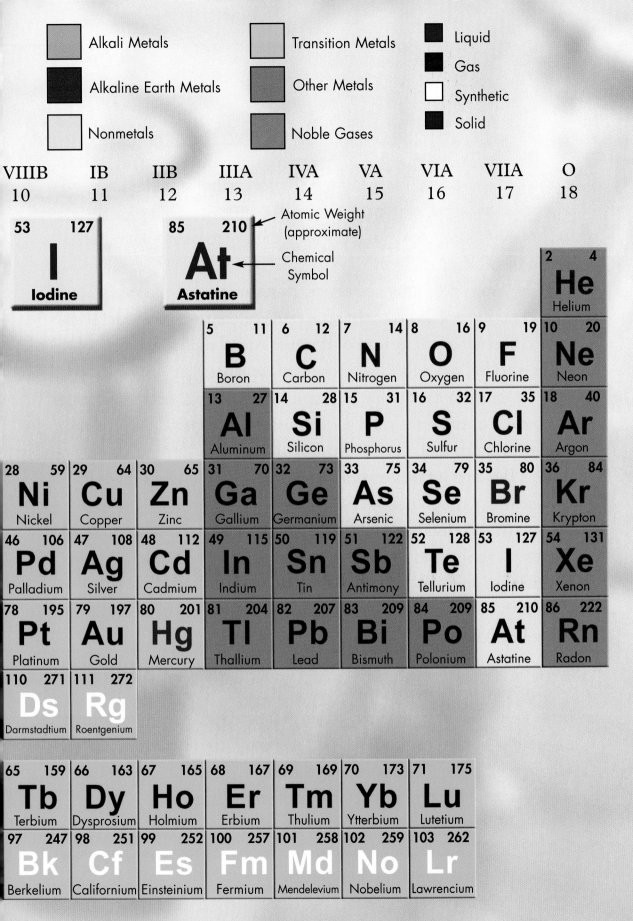

Alkali Metals

Alkaline Earth Metals

Nonmetals

Transition Metals

Other Metals

Noble Gases

Liquid

Gas

Synthetic

Solid

VIIIB	IB	IIB	IIIA	IVA	VA	VIA	VIIA	O
10	11	12	13	14	15	16	17	18

53 127 **I** Iodine

85 210 **At** Astatine

Atomic Weight (approximate)

Chemical Symbol

2 4 **He** Helium

5 11 **B** Boron

6 12 **C** Carbon

7 14 **N** Nitrogen

8 16 **O** Oxygen

9 19 **F** Fluorine

10 20 **Ne** Neon

13 27 **Al** Aluminum

14 28 **Si** Silicon

15 31 **P** Phosphorus

16 32 **S** Sulfur

17 35 **Cl** Chlorine

18 40 **Ar** Argon

28 59 **Ni** Nickel

29 64 **Cu** Copper

30 65 **Zn** Zinc

31 70 **Ga** Gallium

32 73 **Ge** Germanium

33 75 **As** Arsenic

34 79 **Se** Selenium

35 80 **Br** Bromine

36 84 **Kr** Krypton

46 106 **Pd** Palladium

47 108 **Ag** Silver

48 112 **Cd** Cadmium

49 115 **In** Indium

50 119 **Sn** Tin

51 122 **Sb** Antimony

52 128 **Te** Tellurium

53 127 **I** Iodine

54 131 **Xe** Xenon

78 195 **Pt** Platinum

79 197 **Au** Gold

80 201 **Hg** Mercury

81 204 **Tl** Thallium

82 207 **Pb** Lead

83 209 **Bi** Bismuth

84 209 **Po** Polonium

85 210 **At** Astatine

86 222 **Rn** Radon

110 271 **Ds** Darmstadtium

111 272 **Rg** Roentgenium

65 159 **Tb** Terbium

66 163 **Dy** Dysprosium

67 165 **Ho** Holmium

68 167 **Er** Erbium

69 169 **Tm** Thulium

70 173 **Yb** Ytterbium

71 175 **Lu** Lutetium

97 247 **Bk** Berkelium

98 251 **Cf** Californium

99 252 **Es** Einsteinium

100 257 **Fm** Fermium

101 258 **Md** Mendelevium

102 259 **No** Nobelium

103 262 **Lr** Lawrencium

Glossary

aerosol can A container holding a substance that is under pressure and released in the form of a spray.

anesthetic A substance that reduces sensitivity to pain, often used during medical procedures.

corrosive Able to destroy something by chemical action.

distill To purify a liquid by heating it and condensing its vapor.

E. coli *Escherichia coli*, a type of bacteria found in the intestines of warm-blooded animals. Some forms cause illness.

enrich To improve the quality of uranium for use in a nuclear reactor.

etch To cut a design in a solid surface using a corrosive, such as an acid, or a sharp tool.

half-life The time it takes a radioactive substance to lose half of its radioactivity through decay.

osteoporosis A disease in which the bones become porous and weak.

polymer A compound whose molecules are made up of long chains of repeating units.

propellant A compressed chemical used to expel the contents of an aerosol container.

radiation therapy The treatment of a disease, such as cancer, with radiation; also called radiotherapy.

radioactive Referring to a substance that emits energy in the form of a stream of subatomic particles and radiation as it decays.

smelt To melt ore and separate metal from it.

subterranean Existing or occurring below the ground.

synthesize To produce something artificially.

thyroid gland An organ located in the neck that secretes chemicals responsible for metabolism and growth.

For More Information

American Chemical Society
1155 Sixteenth Street NW
Washington, DC 20036
(800) 227-5558
Web site: http://www.chemistry.org
This organization provides news, information, and resources for chemists.

American Chemistry Council (ACC)
1300 Wilson Boulevard
Arlington, VA 22209
(703) 741-5000
Web site: http://www.americanchemistry.com
The ACC represents the leading chemical manufacturers in the country
   and is an advocate for the use of chemicals to improve life on Earth.

American Dental Association (ADA)
211 East Chicago Avenue
Chicago, IL 60611-2678
(312) 440-2500
Web site: http://www.ada.org
Founded in 1859, the ADA is the world's oldest professional dental
   association and an advocate of water fluoridation.

Chemical Institute of Canada (CIC)
130 Slater Street, Suite 550
Ottawa, ON KIP 6E2
Canada

(888) 542-2242

Web site: http://www.chemist.ca

The CIC is a professional association of chemists and related professionals dedicated to promoting modern science and chemistry practices.

International Union of Pure and Applied Chemistry (IUPAC)

IUPAC Secretariat

104 T.W. Alexander Drive, Building 19

Research Triangle Park, NC 27709

Web site: http://www.iupac.org

An international and nongovernmental advocate of the chemical sciences, the IUPAC is the recognized authority for the naming of the elements and their compounds.

U.S. Geological Survey (USGS)

12201 Sunrise Valley Drive

Reston, VA 20192

(888) 275-8747

Web site: http://www.usgs.gov

The USGS is a scientific agency of the U.S. government that studies Earth's landscapes and natural resources, as well as natural hazards that put the planet at risk.

Web Sites

Due to the changing nature of Internet links, Rosen Publishing has developed an online list of Web sites related to the subject of this book. This site is updated regularly. Please use this link to access the list:

http://www.rosenlinks.com/uept/halo

For Further Reading

Hasan, Heather. *Fluorine*. New York, NY: Rosen Publishing, 2007.

Lew, Kristi. *Iodine*. New York, NY: Rosen Publishing, 2009.

Newmark, Ann, and Laura Buller. *Chemistry*. New York, NY: DK Children, 2005.

Roza, Greg. *Bromine*. New York, NY: Rosen Publishing, 2007.

Tocci, Salvatore. *Chlorine*. Danbury, CT: Children's Press, 2006.

Wiker, Benjamin D. *The Mystery of the Periodic Table*. Bathgate, ND: Bethlehem Books, 2003.

Bibliography

American Academy of Family Physicians. "Fluoridation of Public Water Supplies." Retrieved January 6, 2009 (http://www.aafp.org/online/en/home/clinical/clinicalrecs/fluoridation.html).

American Dental Association. "Fluoridation Facts." 2005. Retrieved January 10, 2009 (http://www.ada.org/public/topics/fluoride/facts/index.asp).

Bennett, Edward, ed. "The Fluoride Debate." 2001. Retrieved January 6, 2009 (http://www.fluoridedebate.com).

Emsley, John. *Nature's Building Blocks: An A–Z Guide of the Elements*. Oxford, England: Oxford University Press, 2001.

Gray, Leon. *Iodine*. New York, NY: Benchmark Books, 2005.

Greenwood, N. N., and A. Earnshaw. *Chemistry of the Elements*. Oxford, England: Butterworth-Heinemann, 2001.

Jackson, Tom. *Fluorine*. New York, NY: Benchmark Books, 2004.

Knapp, Brian. *Chlorine, Fluorine, Bromine, and Iodine*. Danbury, CT: Grolier Educational, 1996.

Krebs, Robert E. *The History and Use of Our Earth's Chemical Elements*. Westport, CT: Greenwood Press 2006.

Lenntech Water Treatment and Air Purification. "Fluorine." Retrieved January 6, 2009 (http://www.lenntech.com/periodic-chart-elements/F-en.htm).

Meletis, Chris D. "Chloride: The Forgotten Essential Mineral." Mineral Resources International, 2003. Retrieved January 11, 2009 (http://www.eletewater.com/elpdf/chloride_meletis.pdf).

Sen, Paul. "Tiny Finding That Opened New Frontier." BBC News, July 25, 2007. Retrieved January 10, 2009 (http://news.bbc.co.uk/1/hi/sci/tech/6914175.stm).

West, Krista. *Bromine*. New York, NY: Marshall Cavendish Benchmark, 2008.

Index

About the Author

Greg Roza has written and edited educational materials for children for the past nine years. He has a master's degree in English from the State University of New York at Fredonia. Roza has long had an interest in scientific topics and spends much of his spare time tinkering with machines around the house. He lives in Hamburg, New York, with his wife, Abigail, and his three children, Autumn, Lincoln, and Daisy.

Photo Credits

Cover, pp. 1, 15, 17, 18, 40–41 by Tahara Anderson; p. 5 Library of Congress Prints and Photographs Division; pp. 7, 8 © Michael Newman/PhotoEdit, Inc.; p. 10 © Mark Boulton/Photo Researchers, Inc.; p. 11 © Zephyr/ Photo Researchers, Inc.; p. 14 Rischgitz/Hulton Archive/Getty Images; p. 22 © Visuals Unlimited/Corbis; p. 23 Reza/Getty Images; p. 25 Post Street Archives; p. 27 Martin Bernetti/AFP/ Getty Images; p. 29 © Bettmann/Corbis; pp. 31, 32, 34 Shutterstock.com; p. 36 Joe Raedle/Getty Images; p. 38 © SGO/Image Point FR/Corbis.

Designer: Tahara Anderson; Photo Researcher: Cindy Reiman